AMA'S ZOO

AMA'S ZOO

by

Ama Hannah

Illustrated & Edited by DearUju

First Edition

Printed in the United States of America

ISBN: 979-8-9990740-4-1

Cover design by: DearUju
Illustration and Edits by: DearUju
Publisher: DearUju

For permissions, licensing inquiries, or bulk orders, please contact:
dearuju2017@gmail.com

I dedicate this book to my mother, who always pushes me to do my best.
She wants me to live the life she couldn't have.
And I will.

AUTHOR'S NOTE

Hi!
I'm Ama and I had so much fun writing this book.

I hope you love reading about animals as much as I love learning and writing about them.
I even added jokes, fun facts, and surprises for you!

Enjoy the adventure!

Table of Contents

AMA'S ZOO
Hello there....
WELCOME.

My name is Ama
and
I will be your guide
for today.
Here's your pass

We have exotic animals, wild animals, and cute animals!

**But ...
be careful, some bite.**

So have fun!

And remember, don't feed them!

Welcome

ZOO MAP

BIRDS

People have been using birds for many years. One type of bird is a chicken. People have kept birds for meat, eggs and pets! Today, people keep chickens on farms.

When the chickens lay eggs, the farmers collect them and put them in the fridge so the eggs don't hatch.

Adult chickens sit on their eggs until the eggs hatch.

A baby chicken is called a chick.

BEARS

Now we have to be quiet.

The bear is sleeping.

But I bet you didn't know that bears are super smart and have big brains for their size!

They can also run really fast, up to 30 miles per hour! That's faster than a human!

Bears also have an amazing sense of smell. They can smell things from miles away...maybe even me!

Wow! I'm so happy we got away from that bear in time!

TIGERS

Oh look! The adult tiger is cleaning its baby.

A baby tiger is called a cub.

Tigers are canivores, that means their main diet is meat.

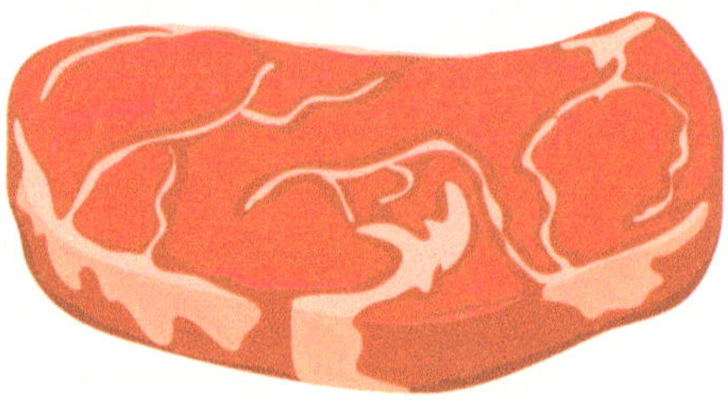

But this depends on where they live.
For example,
Tigers in Asia eat deer

Tigers drink water from streams, lakes and other sources. They also get water from the food they eat.

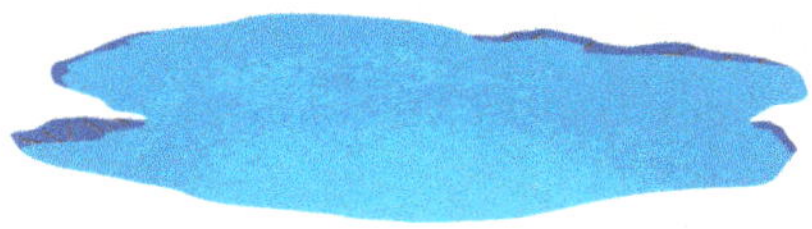

Tigers have all different patterns on their body. Tigers have up to 100 stripes on their whole body! Every tiger has a unique pattern! Pretty cool!

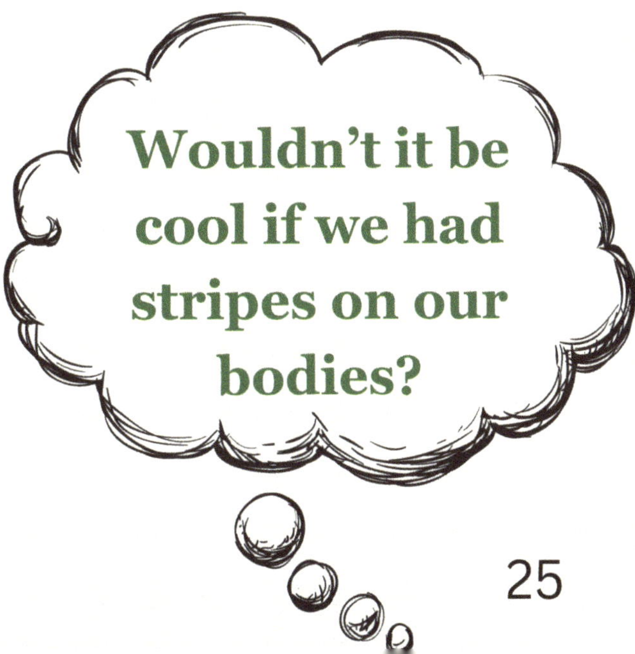

RABBITS

Did you know that the word "bunny" is just a cute name for rabbits?

So yes, bunnies are rabbits! People usually say "bunny" when talking about baby rabbits or when they look especially fluffy and adorable.

Rabbits can hop super fast. Some can leap up to 10 feet in one jump! Their long legs help them run away from danger.

And those big floppy ears? They're not just for hearing. Rabbits use their ears to stay cool in hot places. That's why jackrabbits, which are really speedy hares, live in deserts, where the sun is extra strong.

JackRabbit

Rabbit or Bunny

Uh-oh... that bunny isn't looking. Should we try to chase it? Just to see how fast it can go?

And... GO!

Wait! Come back! That was way too fast!

PANDAS

Pandas are really big! They can grow up to 6 feet long! That's taller than me and even my mom!

Pandas weigh between 150 to 280 pounds, and they're super strong.

They could probably break a chair in an instant!

Pandas spend most of their day eating bamboo, lots of it! But they also eat berries, roots, and sometimes fungi.

Pandas are about 99% vegetarian, but the other 1% of their diet includes things like eggs, small animals, and even carrion (that means the meat of animals that are already dead). I know that sounds kinda gross... but to a panda, it's a tasty treat!

SNAKES

I know snakes might seem scary, but they're actually pretty helpful!

For example, snakes eat mice. That might sound gross, but it's actually great, they help control the mouse population.

Can you imagine having too many mice running around your house? Yikes!

I’m not saying you should get a pet snake (unless you're into that), but if you see one outside in your neighborhood, it’s usually best to just let it be. Now, if you ever see a snake inside your house... yeah, it’s probably time to call animal control!

And some snakes like pythons and anacondas, can be super long. Some can grow up to 30 feet long. That's way bigger than me and my mom put together!

HORSES

You might think, “Horses are boring animals. They don’t even do anything!” But that’s totally wrong.

Horses are actually really important, just like snakes!

Every animal has a purpose on Earth, right?

Horses have helped people for thousands of years. They've been used for transportation, farming, and even fun stuff like horseback riding and sports. My brother loves riding horses and guess what? That's transportation and fun!

Also, horses are strong. They can kick super hard and far, some can kick with enough force to reach 6 to 10 feet behind them. (That's way more than my 2-foot kick!)

So yeah... horses? Not boring. Not even close.

OCTOPUS

Octi are fascinating creatures, but there's one octopus I really want to tell you about: the blue-ringed octopus.

They may be small, but don't let their size fool you, they're one of the most dangerous sea animals in the world!

Those glowing blue rings on their bodies aren't just for show. They're a warning: "Back off!" Its bite can release a powerful venom that can be very harmful to humans.

If someone is bitten, don't panic!

Stay calm (as calm as possible!)

Put pressure on the bite

Call for help immediately!

Most bites are rare, and people survive with quick medical help.

But remember: this little octopus may look cool, but it's not a pet!

WOLVES

Wolves are ferocious creatures. But did you know they are the ancestors of... DOGS?!

I mean those cute little furballs that cuddle with you on the couch? Well... YES!

Wolves and dogs share about 99.9% of the same DNA. That's super close! Even though they don't look exactly alike, they're part of the same family. Wolves are wild, and dogs are... well, let's just say a lot more spoiled.

Dogs can grow up to 99 pounds, depending on the breed. But wolves? They can weigh up to 176 pounds and are much stronger.

They have sharp teeth, thick fur, and a loud, wild howl.

Wolves and dogs might seem very different, but deep down, they're more alike than you think.

Just don't forget, wolves came first!

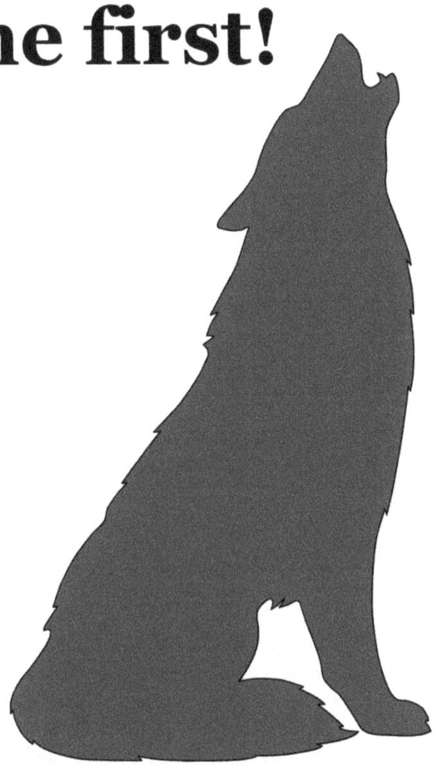

PENGUINS

My favorite animal, and the last one in our adventure, is the emperor penguin!

The emperor penguin is the biggest penguin in the world!

They can grow up to 3.6 to 4.3 feet tall. That's bigger than any other penguin.

They have white bellies, black backs and heads, pale yellow to orange markings around their neck and near their ears and are tall.

Emperor penguins eat krill, fish, and squid - Yum!

Want to hear a cool fact?

They are the only penguins to breed in harsh weather (Antarctic winter)

Breed means they mate and raise babies. Kinda like how some people get married, but in animal form.

I just think
emperor penguins
are so cool.
Don't you?

Well... I guess this is it. Our journey ends here. Thank you so much for not feeding the animals! 🐾

I hope you had an awesome time learning about all kinds of creatures...fast, furry, feathery, and even the ones that almost chased us!

And if you ever see a bear... just make sure it's not that bear...the one that almost caught us!🐻

Until next time, animal adventurer... stay curious and keep exploring!

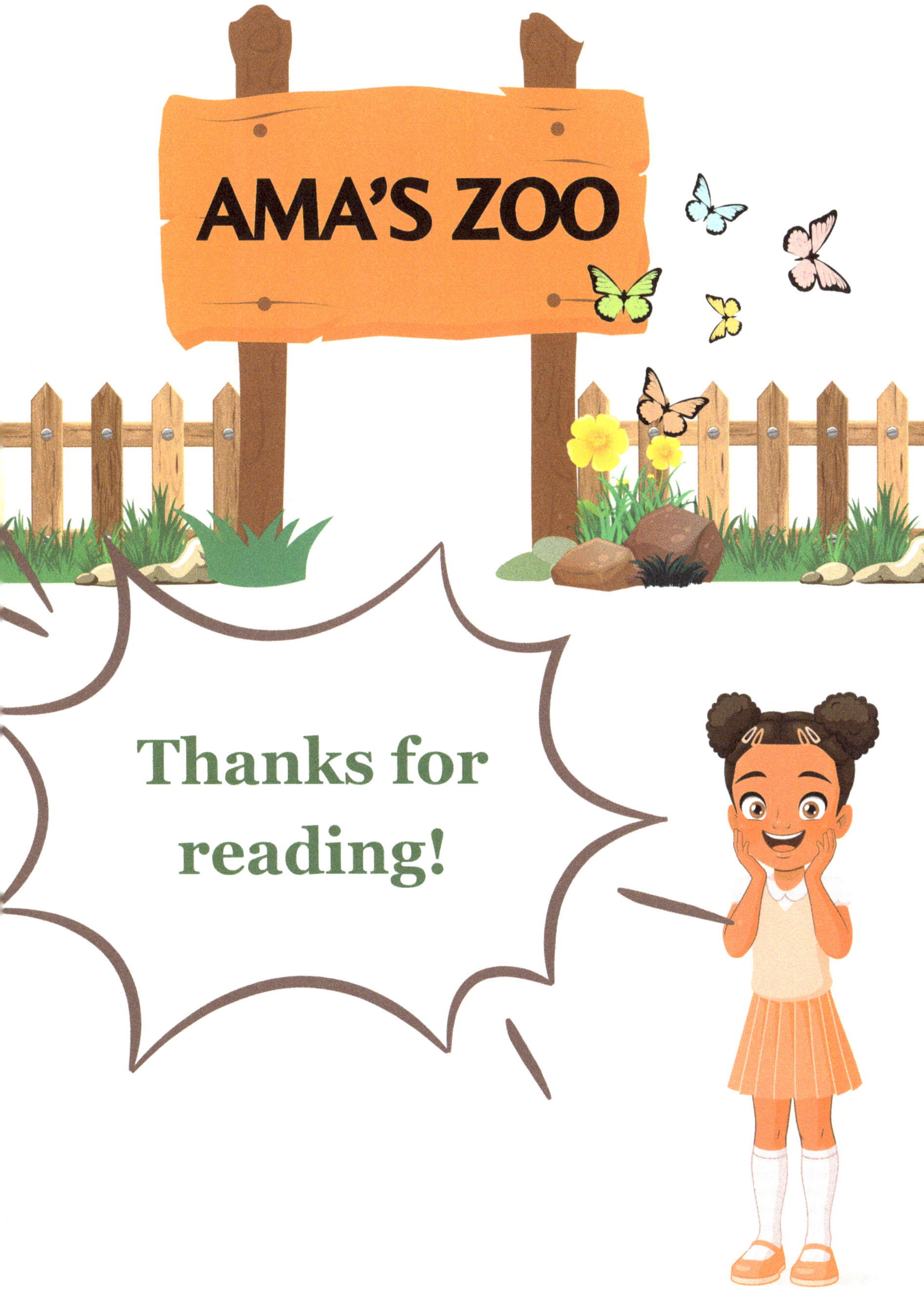
AMA'S ZOO
Thanks for reading!

DID YOU KNOW?

- Chicks start talking to each other while still inside the egg.

- Emperor penguin dads keep the egg warm on their feet for two whole months, without eating.

- Horses can sleep standing up or lying down.

- Wolves don't howl at the moon, they howl to talk to their pack.

- Tigers love water and are strong swimmers. They can swim for miles.

Let’s play!

FUN ACTIVITIES

Draw your favorite Animal

Color with me

Color with me

Color with me

Find the tiger cub

Word Search

M	E	P	A	N	D	A	S	S	S	S	B	S	E
N	Z	V	L	O	T	S	E	N	S	R	S	N	D
P	L	A	N	T	S	B	O	I	P	E	S	O	A
E	S	E	V	L	O	W	M	U	U	B	M	C	R
H	O	R	S	E	S	O	D	G	A	I	D	T	P
O	A	O	N	S	R	E	Z	N	R	A	L	O	S
S	N	A	K	E	S	A	R	E	M	N	H	P	A
E	A	S	T	H	R	W	B	P	E	H	A	U	S
V	G	B	A	K	N	O	E	B	A	R	U	S	R
A	B	E	T	B	I	R	D	S	I	D	A	P	E
E	E	E	P	L	O	A	S	O	O	T	B	E	G
L	M	A	A	N	O	A	N	O	R	L	S	E	I
G	O	R	A	R	A	M	Z	I	E	K	S	R	T
E	H	B	I	U	S	A	E	R	H	T	E	E	S

PLANTS	WOLVES	BEARS	RABBITS	LEAVES
BIRDS	PENGUINS	HOME	PANDAS	AMA
SNAKES	ZOO	TIGERS	OCTOPUS	HORSES

Which Animal From AMA'S ZOO Are You?

Answer these questions to find your animal match!

1. What's your favorite thing to do?

A. Run really fast
B. Climb and swing
C. Swim in cool water
D. Eat snacks all day
E. Take quiet naps
F. Be with my friends

2. If someone surprised you, what would you do?

A. Growl and stand tall
B. Hide in a tree
C. Swim away fast
D. Roll into a ball and chew bamboo
E. Just look at them like... "Really?"
F. Stare until they leave

3. What's your favorite kind of weather?

A. Snowy and cold
B. Hot and sunny
C. Rainy and cloudy
D. Cool and breezy
E. Any weather—if there's food!
F. Cold nights with my pack

4. What color would you wear if you were an animal?

A. Orange with black stripes
B. Brown fur
C. Black and white
D. Gray and fluffy
E. Blue and shiny
F. Gray and white fur

Animal Test Results

- *Mostly A's:* ***You're a Tiger – brave, bold, and love a good snack.***

- *Mostly B's:* ***You're a Bunny – fast, fluffy, and full of curiosity! You're small but mighty and always ready for a race.***

- *Mostly C's:* ***You're an Emperor Penguin – cool, calm, and loyal to your crew.***

- *Mostly D's:* ***You're a Panda – gentle, relaxed, and love quiet snacks.***

- *Mostly E's:* ***You're a Blue-Ringed Octopus – calm on the outside but strong and smart!***

Mostly F's: ***You're a Wolf - strong, protective, and loyal. You might seem calm, but you're always ready to protect those you care about.***

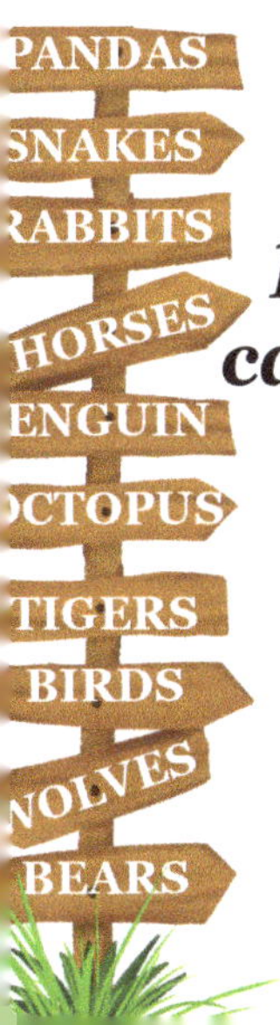

Animal Trivia

(1) Which animal in AMA'S ZOO has blue spots, is small, but very dangerous?

A. Panda

B. Blue-ringed octopus

C. Snake

(2) What percent of a panda's diet is vegetarian?

A. 99%

B. 50%

C. 80%

(3) What animal is the ancestor of dogs?

A. Cat

B. Bear

C. Wolf

(4) What color is an emperor penguin's neck?

A. Orange

B. Blue

C. Red

(5) What's a baby tiger called?

A. Cub

B. Pup

C. Chick

TRIVIA ANSWERS:

(1) B; (2) A; (3) C; (4) A; (5) A

ABOUT THE AUTHOR

Ama is nine years old and lives on the East Coast of the United States with her parents and her baby brother.

She loves animals, riding her bike, and eating pasta! Her favorite subject in school is writing, and her favorite color is light purple.

Ama wrote this book to share some of her favorite animals with you.

She hopes you had as much fun reading it as she had writing it!

AUTHOR CARD

🎈 Name: Amarachi (you can call me Ama!)

🌟 Age: 9 years old

🧡 Meaning of my name: "God's Grace"

🏠 Lives: On the East Coast with my awesome family

🇳🇬 Fun Fact: My parents are from Nigeria!

🐯 Fave animals: Too many to count

🚲 Loves to: Dance, do gymnastics, write stories, read books, ride my bike, play piano

🍝 Fave food: Pasta forever!

💜 Fave color: Light purple

💕 Best thing ever: Family + Friends = ❤️

✏️ Why I wrote this book:

Because I LOVE animals and enjoy learning cool things about them. I wanted to share that love with you and maybe even help you discover a new favorite animal too!

Thanks for visiting my zoo!

Illustrations & design by DearUju

Want more?
Follow Ama’s Journey
on IG @oh_so_Ama

www.ingramcontent.com/pod-product-compliance
Lightning Source LLC
LaVergne TN
LVHW071341210926
848486LV00003B/204
* 9 7 9 8 9 9 9 0 7 4 0 4 1 *